Adaptations 2009

New Science Collection

Adaptations 2009

What if an emotion lasted too long?

Fernando Perez

ISBN 978-0-557-28304-0

/

Dedicated to Monica

Contents

Introduction

This is a book about life, the universe and everything.

An adaptation is the rearrangement of a system due to a new external stimulus, so that a new system arises, one prepared to deal with that external stimulus.

In all adaptations the new system becomes dependant on the external stimulus that changed it. Sometimes that dependence is very low, sometimes it is a total need.

Some adaptations are reciprocal between two systems. This happens when two systems affect each other constantly, and once one of them morphs the other one will be affected and morph as well.

Reciprocal adaptations can lead to chain adaptations, which is what happens when system A morphs and causes system B to morph as well, then the change in system B affects system A and it morphs again, which affects system B again… and so on.

Chain adaptations can have two outcomes: one of them is reaching a sustainable balance when the systems reach a extreme where they don't have room to keep changing anymore.

The other possible outcome is a perpetual chain of changes. We call this an evolution. This is the case with the evolution of the universe, biological life, economy.

Heat

With time, things change. But sometimes a system starts changing faster than the flow of time indicates it should and we say the system is heating up. If changes are coming slower than expected we say that things are cooling down. That's what heat is in a physical level, it is an increase in the speed at which particles are moving inside an object. When an object is cold it means its particles are moving slowly. In the field of chemistry heat accelerates the rate at which reactions happen. Heat is a difference of time flow, a change in the rate at which things are happening.

Evolution

Evolutions in our universe are chains of adaptations, changes, the transformation of simple systems into more complex systems. The level of complexity is always rising.

Every time a new tier of more complex systems arises, the previous ones keep existing, as they are needed to support the new systems, for the more complex a system, the bigger its needs are. The older systems create a layer of stability that the new system needs.

Complexity has been ever rising from the start of the universe.

Most evolutions happen when an adaptation has more uses than it needed to and it gives unexpected advantages that are discovered by accident. The first birds were not expecting to fly. Be Spread was not expected to lead to phant.

Life

When a group of systems arrange to create a new, more complex supersystem, we say a higher lifeform has been created. Life is the composition itself, and when the supersystem decomposes and it's subsystems go free, we say the lifeform died. It applies to companies and abstract systems, we talk about them being born and dying, which implies they have lives.

The passage of time causes things to decompose, to die. Those systems that avoid decomposition better are the ones that remain, survival of the fittest applies to all sort of systems, not just organic life. Evolution and life is a battle against the effects of time.

The Universe

The universe is the space that contains all the energy we know. Energy is constantly being created and building up, which causes the universe to expand from its initial small size to the present, to keep energy density somewhat constant.

Energy has a way to spread around the entire universe. Every star emits spheres of light that expand to the rest of the universe unless they come across something. Every point in this universe is awash with energy flowing from every direction.

Matter

Energy can bind with more energy so that instead of radiating away, it becomes stable, it becomes matter. Matter is produced by energy under the right circumstances.

In the early stages of this universe there we only produced lifeforms, systems that were created by forces whenever the situation was ideal. This was the case with galaxies, planets, water.

The complexity of behavior rises smoothly from the elemental particles to the human society. Psychology is an extension of physics, they are both sciences of behavior.

Some types of energy are solid. Solid means that they push their peers out of their space, they are as territorial as animals can be. What we see as a solid object is actually the action of pushing away anything that gets near. There are no objects, just actions.

Chemistry

On Earth, billions of years ago, a higher lifeform was produced that same way. Organic life as we know it had begun. Organic life is a chemical chain reaction. Reproduction probably appeared much after chemical chain reactions became self sustainable, the first lifeform were probably expecting eternal life instead of reproducing.

Economy

Chemical reactions consume materials constantly, so lifeforms based on chemical chain reactions are always consuming resources. They need resources to sustain the chemistry. The concept of "need" was born. And much later, when lifeforms gained the ability to fight for resources, economy was born.

A side effect of chemical chain reaction lifeforms is that they are always rebuilding themselves from raw materials, therefore they have the ability to repair themselves out of most damage.

Organic lifeforms are also very unstable, they don't live as long as a rock or a lake. Reproduction ensures that the lifeform keeps existing.

Biological Evolution

Organic life quickly evolved into more complex systems thanks to its ability to reproduce into slightly different copies of itself, which sometimes were better, sometimes worse. The better ones remained thanks to "survival of the fittest", which is based on a competition for a limited amount of resources, economic success. If resources are plenty everyone gets to survive, as it is happening to modern humans, and then devolution happens, all sorts of mutations get to survive and reproduce.

Thermoregulation

Evolution always seems to need a more and more stable environment. Only when a more stable environment appears or is built can a higher level of evolution be reached. One of the ways this has happened is with the control of temperature in living beings. More evolved animals can only function in an ever decreasing range of temperatures and this is possible thanks to the ever increasing number of tools the body has available to control its own temperature.

Animals with a small number of such tools/techniques are called ectotherms or cold blooded. These animals have so few tools to control their heat that their temperature is mostly that of their environment.

More evolved beings have enough tools to guarantee their temperature will remain within a small range. In humans the goal is around 37°. The downside of this is that controlling one's temperature is very expensive energetically, so these types of animals have higher needs, they eat a lot more food and have stored reserves that they burn whenever heat is needed.

DNA

A higher level of complexity was reached when molecules specialized to have different functions inside a lifeform. A DNA molecule took the role of center of command, coordinating many actions around the lifeform by means of starting chemical reactions by sending molecules to start them. The DNA molecule took this role because it was the only one capable of encoding a long list of instructions, the only molecule that can form chains as long as needed.

A DNA molecule is the central part of the cell, but it is meaningless without the engine that supports it and reacts to it, the rest of the cell.

Much later on Earth, more complex beings appeared that were composed of different cells bound together, each one with a different function. Neurons, nerves, skin cells, blood cells. A higher level of complexity had been started

Neurons

A special type of cell adapted into being a centre of command of multicellular beings, the neuron and similar types of cells that form the central nervous system. This is a high complexity version of the DNA in cells, the central nervous system is the center of command and the rest of the body is the engine that executes its instructions.

There are many parts to the central nervous system, the most important of which is the reticular formation which acts as a selector between several predefined neurological states, for example, it is in charge of switching the animal from awake to sleep at night and then switching it again to awake in the morning.

The switch between neurological states plays a key role in emotions but also in the definition of a personality.

Neurons implement instincts, which are drivers of behaviors predefined by DNA. They can be switched on or off during changes of neurological states.

Learning

The brain learns by establishing connections between neurons. When we perceive a series of events happening twice, a chain of connections in the brain grows stronger, and if it happens again it grows stronger, until we learn that one event will always follow the other. Most times we are unaware of this learning, and we just start predicting the final events whenever the chain starts, and we can't explain why, we just blame it on a sixth sense.

A well filled up brain contains complex models of the world around it, which can be used to test possibilities and imagine results: ideas can be generated. Once an idea is tried and it works, it becomes part of the learned facts itself, and the models keep growing bigger and better.

Learning like this happens on the brain, not on the mind. The mind doesn't learn. When someone learns to play the piano, it is his brain that learned, his consciousness cannot play the piano, his brain and body play almost on autopilot, same happens with sport players. The mind just watches it all happen and can interrupt it, start it, stop it and modify it, but never control its details.

Neurons are a very valuable resource in the brain and can be reassigned to a different task depending on what is on demand. Every time something is learned, neurons were reassigned. Neurons that aren't being used are preferred for reassignment but location is more important. All neurons belonging to related tasks are located together and when a new task is learned, neurons nearby similar tasks are preferred for reassignment. When neurons are reassigned to something else, an old task will be slightly forgotten, from the outside this is what we call "loss of practice"

Habits

The repetition of an activity can lead the brain to learn it and start repeating it automatically. This is many times desirable but will happen even if it isn't needed anymore and serves no purpose. Habits can slowly fade away if the activity is stopped but usually they don't disappear completely unless something opposite is learned, for example if the habit becomes painful. Also habits fade when something better is discovered and a more pleasant habit is built out of it, replacing the previous one.

Once you have been doing something for a while, your body starts developing physical, organic adaptations to better support that activity, like growing muscles after lots of physical exercise. Once that happens, the activity becomes a need, because your body is now built to perform it, so not performing it is a waste of resources that hurts somewhere. The adaptations that make habits so strong don't only happen in the brain, the body adapts too, and the lifestyle of the person, the things he owns, even other people when affected by someone's habit, adapt to it and end up reinforcing it, even if unwillingly.

Imitation

Learning takes time, and it would be shame to see it go to waste when the animals die. After brains evolved to be capable of complex learning a useful adaptation appeared, the instinct to imitate.

One of the most effective ways to learn behaviors that can't be coded in the DNA is imitation. Animals are born with mirror neurons, which cause them to want to imitate the behaviors of others even when no reward is perceived.

This is essential in shaping up cultures and gives cohesion to a society of animals; it causes everyone to do the same things.

When an animal is born without mirror neurons by genetic mutation he misses on a lot of learning and will always be in serious disadvantage if competing with his peers for resources. This is less of an issue in humans were an even more complex way of learning is available: thoughts.

Palm Tissue

Around ten million years ago one of the animal species evolved something special, something unique. The primates adapted to a lifestyle of hanging from tree branches. This adaptation included developing large palms on their hands, with a special skin tissue that is flexible enough to grasp any branch, but strong enough not to tear up.

That was the most beautiful accident in history; as soon as these primates put their palms near their brains they had deeper, longer chains of thoughts. Since childhood they would quickly learn to put their hands near their mouths whenever they needed to think and that's when the hurricane that turned them into humans started turning.

Larger chains of thoughts resulted in better ideas, higher intelligence, and it became their biggest advantage in the fight for survival. With time their bodies started adapting to a peaceful thinker lifestyle, they lost body hair, started walking erect, evolved opposable thumbs for tool manipulation, manipulation of symbols, verbal communication, but more important of all, their brains grew bigger and bigger to contain more learned behaviors and to emit phant of its own. But the palms remained the same and we still use them when complex thoughts are needed.

This type of skin tissue probably existed before in other animals but never in such a large area, in a part of the body that has the flexibility to move further or closer to the brain at will.

The effects of the tissue of the palms of the hands may be the complete explanation for babies sucking on their thumbs, the nail biting when nervous and all other situations where humans put their hands on their mouths. There is also a common posture where the palms go over the back of the head when desperation or worrying is being felt.

After this humans gained the ability to "think about it later", which animals can't do and technological advances started, human lifestyle became more stable and the bodies of people started adapting. The bodies of females adapter faster along a line of changes that includes losing physical strength and body hair. Loss of body hair means a dependence on cloths but also a lower body temperature, which seems to cause a less animal and more human behavior / personality.

Mind

As organic life became more and more complex its needs kept growing, the world was becoming too complex to be navigated with genetically/chemically preprogrammed behaviors. Animals at the time were physically capable of many more things than they were doing, they needed a deeper understanding of the world, a capacity for long term analysis. And eventually the mind evolved... by accident, one of the most beautiful accidents nature has had.

The mind is what sets humans apart from the rest of animals, it is basically a flow of thoughts. It is a tool of the body to be in charge of long term planning. The physical brain lends it all the tools it needs to do so, such as imagination and memory. Ideas are generated by the brain and communicated to the mind. The brain has many internal processes that the mind never hears about. Very little (or a lot, depending on point of view) communication happens between the two. The brain can arrange a complex response to a virus and invest many resources into it, and the mind will never know it happened. Conversely, the mind has little communication to the body, it can have a long chain of thoughts while the body just stands still. The mind can request movements but only on a big scale, it is never given low level control of each muscle, it probably wouldn't be able to handle the load since it can only focus on one thing at once, opposite to the brain who can do many things at once. The brain on the other hand has full access to every single muscle and body function and can organize very complex scenarios.

The mind depends on the brain in many ways. The brain can refuse memories, it can offer ideas out of the blue or not at all. The brain can supply imagination or deny it. You can try to picture a scene in your mind and fail.

Regarding the location of the mind inside the body, there is no record of a person losing his ability to think from injury to any part of the brain and if we consider the amount of communication happening between the brain and the mind all day long it follows that if the mind was located in a certain area of the brain we would see a lot of wiring going in and out of that area. Since no such central area can be seen in the brain the only possible explanation is that the mind is... all over the brain! This would allow the mind to survive minor brain injuries, regardless of their location. Maybe the mind is more like a cloud.

What we usually call "self" is the union of our mind plus the common behaviors we remember doing plus what we can see of our body.

The basic features of the mind are feelings and will. Thoughts happen on terms the brain understand, they usually involve images and sounds that the brain has learned before or the type of symbols and knowledge that the neural network can store. In this sense the mind is like a librarian at a library, going from shelf to shelf picking specific books and using them. The mind is like a cloud all around and inside the brain that gives little ticks in the right places to cause stored information to go from place to place, shaping up what we call a thought.

Theory of mind

During imagination human brains use the assumption that every other system works the same way as them. When applied to non humans this phenomenon is called anthropomorphization and includes thoughts like "the car must be tired after climbing that hill", or "my computer is mad at me".

When applied to humans it is called Theory of Mind and it leads people to believe every other person wants and likes the same things as them. Theory of mind is the automatic assumption that everyone else experiences the world the same way as I do. People who like a certain food will often go around telling anybody "try it, you will love it", ignoring the fact that different people have different tastes.

People who like to be spoiled assume that everyone else likes it too. Those who like to be alone assume that everyone else likes to be alone too.

The process of "getting to know" somebody and realizing their differences defeats theory of mind, but since it is not possible to scientifically know everybody and every system before interacting with them, theory of mind is a very helpful shortcut and social behavior would be impossible without it.

Memory

Memory in organic creatures is stored in the brain. The mind has no memory, if the brain starts feeding it different memories it will react according to them, just as it happens when the mind wakes up during a dream, turning it into a lucid dream. The mind begins participating in the dream, completely unaware of the real world it normally lives on.

The mind thinks it has the ability to relive memories but it doesn't. What happens there is pretty much imagination. Facts are read from the brain's neural networks and then a scene is imagined where all the details are made up on the spot, all characters have the same personality and the viewpoint can be moved to unreal places.

In fact the personality of such characters is the current personality of the person doing the remembering, since the brain is not powerful enough to create other personalities on the fly, the only thing it can do is take the one personality it know very well and reuse it over and over. If the person changes personality, his memories will seem to change as well, the characters in those scenes will have that new personality.

Imagination

This was briefly mentioned before as one of the basic abilities of a brain. It is the ability to represent external objects as internal models and playing with those models, testing possible scenarios.

Now this can be combined with control theory to reveal a very interesting detail: Imagination is also controlled by the chemical/electrical brain, which means it is limited by the same current configuration. In the case of vocal range, people with limits can only imagine their inner monologue voice with that limit too. People who can only do some facial expressions, will imagine only those as well.

You can only imagine that which you can do, because imagination uses the same circuitry real actions use. Trying to imagine something you can't do will result in a shallow scenario with all details missing. Since imagining some action reinforces the same neural networks than actually doing it, turns out that imagination is a good form of training and get better at something.

Predictions

The brain tries to predict and prepare feelings, emotions, sounds, images, smells and tastes. When you approach a mirror your brain is already preparing the image of you it will present. When you taste a drink half of the flavor was predicted by your brain and the other half is the actual input from your tongue.

This helps to keep things constant, it makes the personality and the behavior more constant and predictable.

Intelligence

Intelligence is the ability to predict the outcome of an action. Non-intelligent acts are reactive, they happen as a part of a chain of cause-effect. Intelligent acts are proactive, they are planned in the mind in the shape of ideas. The further into the future a system can predict outcomes, the more intelligent the system is.

Non-regulated collective systems, like early economies, do not show any intelligence, they follow whatever path. Companies on the other hand have always had a central intelligence running them.

The actions of one cannot change the collective result, it is usually a waste of time. Real intelligence in a collective can only be achieved when all subjects can be forced to do something they don't want. This is the case with the brain, which can select the main necessity of the moment and then force the entire body to focus on satisfying it.

Control

A system of control is in place in the central nervous system that enables the body's needs to be listened to by the brain.

The brain is not royalty, designed to just enjoy the ride, it is a tool, designed to bring a better life to the entire body. Being a tool, the brain needs to be controlled, to be given instructions. The body needs to make sure that the brain is working for its best interests and not wandering lazily. Something similar happens to the mind, but it has more freedom to use to body for any private purposes it has, like suicide or sacrifices for political views. The brain also can come up with purposes of its own, like the instinct of curiosity. But it is essential that the needs of the body are listened to by these two.

The method this control is exercised is called sensations. Pain is the main method of control; it aligns the priorities of the mind with the needs of the body. But there are many other methods, tastes for example are meant to teach the mind which foods the body wants. If the body's reserves of sugar deplete, the body sends a "crave" for sugary foods to the brain, who most of the times obeys and looks for sugary foods. This gets a bit more complex when we take into account that the needs of the brain can also be ignored/obeyed by the mind, for example habits.

The mind has the ability to ignore these cravings but it rarely does. Human beings are notorious for noticing the negative long term effects of their actions and still perform them, because their bodies tell them to. The measure of the capacity of the mind to overcome control signals from the body or the brain is called will power.

If the body runs out of water a craving for water is sent to the brain, besides the more physical signals like dry mouth. There are also rewards, like sexual pleasure, which is meant to train the brain, to make it want to do it again. Our bodies train our brains just like we would train a dog, by giving a cookie when he obeys and a smack when he disobeys. This high level of control is necessary due to the high level of control the brain has on the body too. If pain didn't exist we would do a lot of damage to our bodies unknowingly.

Different personalities consume different levels of resources, which causes people to crave different foods, some people will prefer sweet foods, others will enjoy salty foods more. This is more evident in how tastes change in pregnant women.

Minds and brains are only controlled while needs are unmet. When all needs are met the mind and brain are given freedom, which usually results in the brain giving in to its habits or instincts. If the brain is also satisfied it usually results in the mind coming up with long term plans or deep analysis of past memories.

Emotions

Emotions are an assembly of body postures, chemical fluids and phant. The more these things happen, the more intense the emotion feels. They are a coordinated effort of body, brain and mind and any of those can start them, though the mind can only do so indirectly by using techniques that have worked in the past (like playing happy music to cause happiness) or by psychological engineering.

When the mind starts an emotion (like when we set out to read jokes online) the brain understands that an emotion is required and gives the proper orders to the body: put the mouth in laughing postures, release chemicals.

Body Language

One of the tasks of the brain is to set up the body into the most effective posture and movements for the task it is currently performing. This can be interpreted externally by other humans using theory of mind, so they are able to get an idea of what the other person is feeling. Since communication is happening, even if unwillingly, we have come to know these body postures as body language. Note that communication is not the purpose of these postures and unsurprisingly said communication doesn't always go right, many times body language is misread by the other parties, which for example happens a lot to individuals in the omega neurostate.

Examples of body language: The act of thinking is more effective if the palm of the hand is held close to the brain. Aggressive behaviors generate a lot of heat, so the body tends to spread out during anger, to dissipate it. Chemical reactions in the body can only happen in a narrow range of temperatures so when external temperature goes below the minimum the body bends inwards to preserve the heat it's producing.

The reverse engineering of body language is the research into the reasons why each body posture is used.

Configuration

There are many things that can be controlled in a human body and the brain has access to the configuration of all of them. The brain can make it so that you feel an awful flavor when you taste chocolate, it can make sexual pleasure go away, it can make you want to bend into a fetal position when sleeping, it can make you laugh at different things than everyone else. There are so many variables to be configured that the combinations are endless and this is the base to the diversity of personalities. A human personality is the configuration of all those variables.

The choosing of a value for any of the variables brings a limit to the range of emotions the subject can feel. For example, if the subject is configured to be bent inwards most the time he will never experience full rage. All configurations bring benefits and limits.

How the brain chooses a configuration depends on many factors, of which the most relevant is the case of an emotion lasting too long which leads into neurostates.

Voice

One of the things that can be configured by the brain is the vocal range.

Human voice is produced by vibrations in the throat that resonate across the chest or the skull. Due to their size, the chest produces low tones and the skull produces high tones. The human then learns how to manipulate them both to produce a smooth range of tones.

However, the brain can block access to one of these resonating boxes, or both, resulting in four possible configurations:

1. Chest blocked, skull available. On this configuration the subject will only be able to produce high tones of voice, resulting in what we call a nasal voice. A narrow range of tones are available and they are all higher than normal. This restriction of the vocal range implies that the subject is also restricted in the range of emotions he can experience.

2. Chest available, skull available. On this configuration the subject has all tones available, which makes it easy to convey emotions thru the voice and thus to feel emotions intensely.

3. Chest available, skull blocked. This configuration causes the subject to be only able to produce low tones of voice. This restricts the ability to convey emotions thru changes of tone of voice. The range

of emotions the subject can feel is limited by this. This is the case in the omega neurostate.

4. Chest blocked, skull blocked. This causes the subject to be non-verbal. Subjects find a sort of escape by generating noises thru their nose and using those noises to communicate very basic feelings and ideas. A very limited vocabulary arises with only a handful of noises representing a handful of meanings each. One noise will mean "yes", other will mean "no", other will express surprise, other will mean "I don't care", other will mean "that was funny", other noise will mean "I'm uncomfortable". The subject will still be able to scream though.

Neurostates

Neurological states are the most basic configuration setting in the central nervous system. Animals can only be in one neurostate at a time and up to the development of psychological engineering, neurostates were forever, chosen only once during the early weeks of life.

Basically, when an emotion is felt too often the body adapts to be feeling it all the time as the body language of that emotion becomes a habit and the body starts needing that emotion, starts automatically causing a shallow version of it by habit and this is called being in a neurological state, stuck in the shadow of an emotion.

A neurostate is an adaptation
of the central nervous system to an
emotion that lasted too long.

A neurostate is always at the core of a personality. Habits, opinions, all adapt to the current neurostate.

Among other things a neurostate defines the way food tastes, the feelings of the skin, all five senses in general, the way the person experiences the world, the vocal ranges, the favorite colors. Also the body type is affected by the neurostate, including the size of internal organs, the position of the jaw, the way teeth grow.

During puberty the body will grow into a certain type depending of the neurostate. These changes that are usually considered part of puberty are actually temporary and will change whenever the person changes from one neurostate to another.

- **Alpha:**

Being in alpha neurostate causes the person to be calm, in control, somewhat insensitive to input from all senses, muscles are used to the maximum in every movement causing the impression of strength, empathy is high causing the person to want to make everyone happy. Pain tolerance is high. The person is impulsive, likes to just try things without thinking them properly. Highly resistant to the common cold and allergies. Body is wide.

- **Beta:**

People in beta neurostate find it difficult to fall in love.

- **Sigma:**

The sigma neurostate causes the person to obsess on physical pleasures, especially food flavors. Also enjoy the attention of others, to the point of seeing them as servants and using them like tools. Pain tolerance is very low. Comfort is a need and efforts are minimized.

- **Omega:**

In the omega neurostate mental activity is highest, leading to the development of abilities like a vivid imagination rich in details. All five senses are hypersensitive to the point of causing discomfort on certain inputs. Core body temperature is low which causes a propensity to flu. Propensity to allergies. Empathy is low to the point of enjoy fantasies involving other people in pain. The person usually spends a lot of time lost in his own thoughts and gets upset at interruptions because it takes a while to get so deep into his thoughts again. Poor body control, unable to dance properly. Excessive planning of every little detail before actually doing something. Voice is of the type 3 as described earlier.

Conditions

There are conditions of the nervous system, of the brain and of the mind.

Neurological conditions are those that imply a deterioration of the biological processes that make up the nervous system, as is the case in Parkinson and Alzheimer.

Brain conditions arise when the complex processes of a human brain fail and produce all sorts of odd results such as hearing voices, seeing ghosts, hallucinations, memory loss. The neurons are usually healthy in such cases.

The mind itself is very simple and thus rarely sick. It's only malfunctions are the condition underlying what is commonly identified as evil eye and the hypochondria, which is the induction of physical symptoms of any disease by the mind itself. In both cases the brain is physically healthy.

On the other hand there is a popular tendency to call mental conditions to any behavior that is inconvenient to the majority of the population. Under this view a person's mental health is relative to time, situation and to the person making the assessment. Any difference will be called sickness by those in the current majority. This allows for neurostates to be called mental conditions.

Locks

Since brain processes are affected by body postures and activities, and body postures and activities are bound to become habits, it happens that sometimes the body acquires habits that lock it into a neurostate. Bending inwards into fetal position constantly locks the person into omega neurostate. If the person starts spreading a bit the omega neurostate breaks. Holding the hand near the mouth can also become a habit and force the person to spend a lot of time thinking.

Personalities

The personality is the causes of the behavior of a person. It is developed thru life, as opposed to the body structure which is determined since conception by the DNA. A personality is an adaptation to the situations the person has had to face. Different personalities can arise from a same DNA. However the personality is defined entirely by the brain and not by the mind. The mind is mostly a spectator to the personality, it perceives the personality to be unavoidable.

DNA is setup up so that a set of different personalities will rise in a group that will complement each other. Collective behavior is part of the evolution of DNA.

Personalities tend to last all lifetime but they can change naturally or thru psychological engineering.

The personality is made up of habits, learned facts, genetic limitations and a neurostate.

The brain seems to recognize the environment and change personality accordingly. We use different behaviors in an office, in a stadium, in a party. Even the cloths we wear can change our behavior.

Psychological Engineering

Psychological Engineering is the application of techniques to switch the neurostate of a living being.

Those techniques are provided by the reverse engineering of body language or by experiments. A lot of work in this field was done thousands of years ago during the creation of yoga.

It is focused on the creation of physical habits that work as a lock into a new personality type, escaping the previous one.

The reverse engineering of human body language started in 2006 and forms the base of psychological engineering.

Psychological engineering should open the door to more diversity in the personalities of human being. Technology has been changing fast and humans haven't been able to adapt properly as they had no control over their personalities, they just became more and more dependent on being served and lost skills in the process.

Lack of Mirror Neurons

A lack of mirror neurons is present in many people. It is hereditary to 50% of the offspring if only one of the parents has it. It behaves genetically as a dominant trait and it is four times more common in males than females.

It has many implications. For one, it could lead to a cleansing of unnecessary behaviors that have been preserved and passed on by people with mirror neurons when they aren't needed anymore. A proliferation of people with this trait could cause a quicker adaptation to new environment. People with this trait have an easy time seeing things from a new perspective.

The main side effect of it is an isolation of the person from society, since he doesn't learn all sorts of manners and customs. This leads to trigger events since the first months of life.

Trigger Events

When somebody with a lack of mirror neurons faces a social situation where no natural needs are being serviced, something that was learned by everyone by imitation, the person sinks into an emotion of rejectedness and inferiority, an emotion slightly like anxiety. If this emotion lasts too long, as is the case in someone who will go his entire life without imitating blindly, habits are developed and the person is locked into the omega neurostate.

During these trigger events the body tries to get into fetal position and the temperature of the front of the brain drops. As the person enters omega neurostate after a series of trigger events this lower temperature expands to the rest of the brain.

ONS

The Omega NeuroState was first researched in Vienna in the 1930s and has the following symptoms:

- Tendency to bend inwards, bending the knees, keeping legs closed or crossed, keeping the hands together, keeping the hands closed, putting one or both hands between the legs. It's as if every part of the body wants to be as close as possible to any other part of the body, and the mind experiences a nice feeling whenever a part of the body is pressed like that. An interesting side effect of this is that body heat is preserved better, which leads to the body making a smaller effort to heat itself up. Core body temperature tends to be lower when in the omega neurostate. As an effect of this, the flu virus is harder to eradicate as it feels right at home in the lower temperatures, so the common cold lasts much longer. Cold temperatures do not feel as uncomfortable as in other neurostates. Finally, another side effect of the lower core temperatures is slower aging. People who spent decades in this neurostate tend to look years younger than their age. In relation to the temperature issue, symptoms are known to fade a bit during fevers, especially on young kids whose behavior is still not driven by habits.

- Daydreaming, constantly lost in inner thoughts. As an effect of this the subject experiences bursts of energy that find release by quick movements like clapping or flapping the hands. Also this greatly exercises the brain's ability to picture things leading to the ability of seeing very detailed images in the mind.

- Perhaps related to that, there's a lack of social capacities. The brain doesn't come up with things to say during conversation. Also the ability to read other people's emotions is lessened. This means the brain never learns to predict emotions, which is evidenced during early years when the kid doesn't play games that involve representation of emotions in toys.

- Inability to establishing a phant connection with other person during conversation. They may be looking at each other's eyes, but not "into" each other's eyes.

- Allergies are common in this neurostate and they fade away during a switch to another neurostate.

- Obsessions are common, seeing something interesting can cause the subject to obsess on it for years.

- Tendency to establish routines, to avoid trying something new.

- The vocal range is limited to low tones as explained in the Voice section.

Will

Will is a main feature of the mind, along with thoughts and feelings.

Human beings can use their will power to mask behavioral traits of their bodies up to the point of building habits that replace their automatic behaviors. Artificial habits take about three months to establish.

It is very common for people with a lack of conversational skills to make big efforts to be more social, speaking about anything that comes to their minds with no direction or purpose other than avoiding silence. This is the way they adapt to being in a world were social contact is necessary to gain access to many things.

These artificial conversations are often deprived of emotional content other than laughs. Other people may sense the lack of emotional content and grow uninterested in the conversation and walk away at their first chance. This is called killing the conversation.

It is heavy work on the brain to work on this level of acting and people who do this may feel like they never quite get there despite constant effort. This acting also causes emotional displays at the wrong time, taking the others by surprise.

On extreme cases the mind gets to the point of implanting made up facts into the brain, when someone convinces himself of something he wants to believe.

The existence of will opens up the possibility of changing the body according to long term plans. Psychological engineering becomes possible. Things like Be Spread.

Be Spread

There are two very important differences between human beings and animals:

- The first is the complex set of postures the human body can use, thanks to the high flexibility of the limbs. This is thanks to our past as branch graspers.

- The other is the ratio of mass between the limbs and the main body. In most animals the core of the body is where all the mass is while the limbs are either short, like in alligators, or too thin and boney, like in spiders and giraffes. This means that in animals when it comes to controlling the levels of energy inside the body, the limbs are not relevant. Most animals are as spherical as possible, as this is the most convenient shape when it comes to manage energy distribution and conservation. For example even if a cow could put two of its paws together, the amount of heat that would build up in that area is nothing compared to the amount of heat stored inside the main body and it would not change the average body temperature in any noticeable way.

What these two differences mean is that human beings are the only ones that can manipulate their energetic configuration just by changing their posture. We can change our geometry, we can manipulate our effective surface to volume ratio, we can go from a ball to a spread shape, and the way energy flows depends a lot on the geometry of the body. A human being in full fetal position builds up heat fast, while the full Be Spread posture will maximize the release of heat.

There are more forms of energy that are affected by the geometry of the body, but heat is the most evident. When it comes to managing heat, the body has a set of tools it can use to keep it steady. This is called thermoregulation. The main feature of thermoregulation is that only the global temperature of the body can be controlled. This means that when a part of the body gets colder, it's not possible to heat up only that part. Only the average temperature is controlled.

Using a fetal position causes heat to build up all over the limbs and body. This causes thermoregulation to cool down the entire body.

Using Be Spread causes heat to be released constantly. This causes thermoregulation to pump more heat to make it up.

On any of the two extremes and the points in between, the head is always in the same posture, which means the amount of heat that escapes thru the head is constant, so whenever thermoregulation is working to cool down the rest of the body it lowers the amount of heat sent to the head, and we end up with a colder brain. Whenever thermoregulation is working to create more heat all over the body it send more heat to the head, which isn't releasing more heat than usual, so it builds up there and we end up with a hotter brain. The difference between the two extremes after the body has adapted to them is around 1° F of brain temperature, enough to cause noticeable changes in the behaviors it creates.

Be Spread interferes with emotional behavior, it doesn't let the body feel weak emotions, while it intensifies strong emotions. This has been studied before, in the field of yoga, certain postures of the body can limit the spectrum of emotions the person can feel.

These changes on the spreading of the limbs are naturally used by the brain for two energy related situations:

- Thermoregulation can also be behavioral. Animals and humans will seek refuge when it's cold and seek the sun when it's hot. Whenever possible, the posture of the body is also used to thermoregulate, for example birds can spread their wings and change their thermodynamic configuration. In humans, the fetal posture is used when it's cold and temperature becomes a precious resource to be conserved. Spreading is used whenever it's hot and the internally produced heat needs to be released as much as possible. This leads to different personality types available to people depending on the climate they live in. Those in warm tropical lands tend to have stronger, aggressive, individualistic personalities. Those in cold latitudes will show more passive tendencies and a sense of working with the community.

- Emotions. Each emotion leads to a certain degree of spreading or bending. Laughing for example pulls the body inwards a little bit. Seeing a cute animal pulls harder inwards. Extreme fear or pain lead to a full fetal position, while anger leads to a fully spread posture. While performing Be Spread the person will notice their body pulling inwards whenever it wants to feel certain emotions.

Now putting these two situations together this question comes up: why is a thermoregulative behavior such a big part of emotions? Because emotions are made up of energy with heat-like properties and because emotions release heat. Anger for example causes a quick heating up of the brain and probably the entire body and this heat needs to be released before it builds up to levels that are harmful to the chemical reactions that make up life. Birds for example tend to ruffle their feathers when angry or aggressive. This releases heat while also giving the impression of a larger size, which is helpful in intimidating enemies.

The importance of temperature for the chemical reactions in our brain can be clearly seen in the yawn, which is meant to bring fresh air into the mouth, to cool down the head as this will keep the person awake.

There is a full spectrum between being in a fetal posture and being fully spread. It is the average of heat released during the day what has an impact in the neurological state of the person. In theory, being in fetal posture half of the day and being fully spread the other half will yield the same result than being in a mid posture the full day.

There are three main cases of human postures and Be Spread can be used on all. These are standing, sitting and lying down.

- Standing up: When standing up some space can be put between the arms and the body as well as separating the feet a little. Things to be avoided are crossing the arms or placing the hands inside the pockets.

- Sitting down: Being spread when sitting down looks a lot like a king in his throne, legs are wide open, arms far from the body, knees bended about 30 degrees, feet resting flat over the floor.

- Lying down: When laying in bed the body can be spread completely, with a 90 degree angle between the legs and 45 degrees between each arm and the body. Hands will rest with the palms facing down to help keep the fingers from curling in.

On all cases the hands can be spread as well, putting some space between the fingers, enough to let air flow between them, but not so much that it hurts, considering that this has to be done for a long time.

Of the three cases, it's the sitting down posture that helps the most as it is used in social situations, when personality is reshaped, although in theory it is the lying down posture that spreads the body the most.

Since cold environments lead to the use of bent postures, Be Spread must be used when the temperature is high enough, which means that extra clothes will have to be used to warm up the body whenever needed. Drinking and eating hot things should be helpful as well.

Being spread is the lock to the alpha neurostate, so people in omega neurostate can psychologically engineer themselves by using Be Spread and enter the alpha neurostate. Maybe any other neurostate can be changed to alpha using Be Spread.

Adaptations

The journey from omega neurostate to alpha thru the use of Be Spread is a complex set of adaptations. There are different timeframes of changes (assuming Be Spread is performed constantly, which may not be easy to get done):

- Immediately: As soon as the body is spread out a sensation of strength can be felt.

- After three days: Mild changes will appear including a sense of relaxation and the brain will be more cooperative when it comes to having ideas for things to say in conversation.

- After a week: The main symptoms of ONS will disappear along the second week. This includes a release of the voice to a full range of tones, an improved body control, core temperature will increase leading to a higher resistance to the flu, obsessions will fade away,.

- After two weeks: The process of adaptation begins. Allergies will disappear, routines will lose importance, bones will reposition. These are longer term changes, they are triggered by begin spread for long periods of time and can only be stopped or reversed with long term bending inwards.

- Habits: Most habits that were built upon features of the previous neurostate will end up being rebuilt. When a person enters a new neurostate for the first time, all it's features will be unknown so the person will have no idea what his preferences should be. With time those new preferences will be discovered and new habits built around them. Of course the raising of new habits depends a lot on the activities undertaken at the time, trying new things helps a lot to make this phase go faster, sticking to the old behaviors can make this never end.

As long as habits from the original neurostate are still there, stopping the Be Spread can eventually lead back to the original personality. Once habits have been replaced the trip back can be harder to achieve.

Grammar

The choice of words and the ordering of them into sentences is not the same in all humans and it may be dependent on the current neurostate.

Shared Personalities

Different cities and countries tend to have certain shared personality traits among their people. Factors for this are the temperature. Different shared personalities lead to different body developments, which cause for example the different accents around the world and the different accents of people from different economic status even if they live in the same place.

Time and Personalities

Personalities today tend to be weaker and spoiled when compared to the hardworking personalities thru the history of mankind.

Singularity

The rates of changes modern society is seeing has been increasing constantly since it came to be. This affects all areas, from economy to technological inventions, to society and personalities.

As more technologies are discovered, they form the base for even more new technologies and the rate of discoveries keeps ever increasing.

By mid twentieth century changes in modern society started happening so fast that a generational breach was created. New generations were now living very different lives than their parents, and they tend to identify with their age group only, leading to youth gangs.

By the early twenty-first century changes are happening so fast that we are using a big part of our time just keeping up.

The 2008/2009 economic recession ended faster than it should and that means the next recession is going to start faster than it should and will keep getting faster until the economic indicators bounce up and down in chaos.

If heat is an increase in the rate of changes in a system, then humanity is being boiled alive in modern society.

Phant

What is phant? I don't know, but it is the explanation behind the behavior seen in the picture on the back of this book and I will explore it deeper on my next book: *Phantoms 2010.*

Written between December 2008 and December 2009.

Bibliography

- Doidge, Norman (2007). *The Brain That Changes Itself: Stories of Personal Triumph from the Frontiers of Brain Science* (1st ed.). New York: Viking Penguin.

About the author

Not much is known. Born in 1,981, earned a degree in Systems Engineering which explains the systemic views behind this book.

From 2,003 to 2,006 a failure in his ability to learn led him to the development of a method of doing research by using the brain as a tool. It starts by avoiding scientific knowledge and looking for raw data instead. Once the brain has enough data inside, it will find patterns on its own, because it is designed to do just that, it only needs to spend time thinking. This is quick research based on light bulb moments, as opposed to the slow pace of the scientific method. Knowledge discovered in this way cannot be constrained to a single subject, it is holistic by nature.

On early June 2005 Monica came to me at a funeral and started telling me details about what was then a mysterious condition called autism. The method had found its first major application. But it's perfect application was found in 2009 in phant since there was no previous research on it and the method doesn't depend on previous research.

Mah Research Pwns!